William Henry Fox Talbot (1800–1877) *Photograph by John Moffat of Edinburgh. May 1864*

A Science Museum Monograph

The First Negatives

An account of the

discovery and early use

of the negative-positive

photographic process

by D. B. Thomas B.Sc, Ph.D

London : Her Majesty's Stationery Office

Foreword

In 1937 the Science Museum acquired from Fox Talbot's grand-daughter the most important collection of photographic incunabula in existence. The collection included the earliest extant negative, a faded and somewhat out-of-focus image on a scrap of paper little larger than a postage stamp, as well as photographs which stand comparison with the finest ever taken. The following monograph prepared by Dr David B. Thomas, Assistant Keeper in the Museum's Department of Chemistry, is based on photographs and documents from this collection.

April 1964

D. H. Follett
Director

Introduction

In 1937 the first negatives and thousands of the very earliest of photographs still lay virtually unseen for ninety years in a cupboard at Lacock Abbey. Later in that year Miss Matilda Talbot, C.B.E., grand-daughter of William Henry Fox Talbot, the inventor of the negative-positive photographic process, generously presented to the Science Museum much of this invaluable material together with some of her grandfather's apparatus, laboratory notebooks and letters. This monograph gives an opportunity for publishing a small selection of the photographs illustrating an account of the discovery and use of the calotype process, the first photographic process to utilize the chemical development of the latent image produced by briefly exposing silver salts to light.

1 Lacock Abbey in 1844
A calotype from
Pencil of Nature

2 Right: Camera obscura used by Fox Talbot

3 Below: One of Talbot's sketches

Villa Melzi
5th Oct. 1833

1 Photogenic drawings

William Henry Fox Talbot (1800–1877) was an only child of upper-class parents He was educated at Harrow and Trinity College, Cambridge, and published his first scientific paper ('On the properties of a certain curve derived from the equilateral hyperbola') at the age of 22. By 1833 he had a dozen papers to his name, mainly on mathematics and optics. He was interested in the application of science as well as pure scientific research and in 1833 published a paper containing a 'proposed method of ascertaining the greatest depth of the ocean'. He was elected to the Fellowship of the Royal Society in 1831. Outside science his interests were philology, theology and politics. In 1832 he became the Liberal Member of Parliament for Chippenham, sitting in the first Parliament after the 1832 Reform Act. But he was not fond of public life and his interest in politics apparently soon died.

He made several extended tours of the Continent, as was usual for a young man of his breeding, and it was on one of these tours in 1833 that he conceived the idea of photography. In order to obtain sketches of views he used an instrument known as the camera obscura (figure 2), a device which had been known in various forms for centuries. A lens at the front of the camera obscura produces an image, which is reflected by an inclined mirror inside the body of the instrument on to thin paper supported by glass. The image formed on the paper is erect (not upside down) and so the outline of the scene may be sketched. The instrument is essentially the same as the viewfinder half of a present day twin-lens reflex camera.

Several years later Talbot described how the use of the camera obscura gave him the idea of photography:

> 'And this led me to reflect on the inimitable beauty of the pictures of nature's painting which the glass lens of the Camera throws upon the paper in its focus —fairy pictures, creatures of a moment, and destined as rapidly to fade away. 'It was during these thoughts that the idea occurred to me . . . how charming it would be if it were possible to cause these natural images to imprint themselves durably, and remain fixed upon the paper!'*

*Talbot, *Pencil of Nature* (1844)

On returning to England Talbot started to experiment with silver salts which he knew from his scientific readings were darkened by light. Unknown to him both Thomas Wedgwood and Nicéphore Niépce had already had the same idea but had abandoned hope of producing fixed camera photographs by means of silver salts. The silver salts darkened too slowly and were difficult to stabilize to light after an image was formed. Yet in 1834 Talbot began to produce fixed camera images on paper. As light caused a darkening of the silver salts the results were negatives. The earliest negative in existence (figure 5) shows one of the diamond-paned windows in the South Gallery of his country house, Lacock Abbey. The paper negative, about one inch square, is lilac in colour and dated August 1835.

In his first experiments Talbot used paper coated with silver nitrate solution, but finding the image too slow in forming he tried treating the silver nitrate paper with a solution of common salt. When the common salt was in excess (giving silver chloride on the paper) he found that the paper was not blackened by light. But when the nitrate was in excess (giving a mixture of silver nitrate and silver chloride on the paper) the sensitivity was sufficient to produce a negative image in the camera in one hour or so. The image was then fixed by treating the paper with an excess of common salt solution, which converted any nitrate which remained into the chloride. It may seem strange that his photogenic drawings, as he called them, could be fixed at all by converting the silver nitrate into the halide salts which even to-day are used as the basis of photographic emulsions. But in fact silver halides can be kept almost indefinitely in the dry, solid state without blackening. However, one result of the use of this method of fixing was the absence of pure whites.

Talbot described how he took these 1835 negatives as follows:

> 'Not having with me in the country a *camera obscura* of any considerable size, I constructed one out of a large box, the image being thrown upon one end of it by a good object-glass fixed at the opposite end. This apparatus being armed with a sensitive paper, was taken out in a summer afternoon, and placed about one hundred yards from a building favourably illuminated by the sun. An hour or two afterwards I opened the box and I found depicted upon the paper a very distinct representation of the building, with the exception of those parts of it which lay in the shade. A little experience in this branch of the art showed me that with a smaller *camera obscura* the effect would be produced in a smaller time. Accordingly I had several small boxes made [figure 4], in which I fixed lenses of shorter focus, and with these I obtained very perfect, but extremely small pictures, such as without great stretch of the imagination might be supposed to be the work of some Lilliputian artist. They require, indeed, examination with a lens to discover all their minutiae.'*

*Tissandier, *A History and Handbook of Photography* Appendix A, 355 (1878)

Sometime between 1835 and 1839 Talbot began to produce positive photogenic drawings by placing the paper negative on top of a freshly sensitized sheet and exposing the new sheet to light transmitted through the negative.

Until 1839 Talbot seems to have made little progress with his photogenic drawing process and may, in fact, have spent very little time on it. Between 1834 and 1839 he published a further eleven scientific papers on such varied subjects as the nature

4 Above: Three small cameras used by Talbot, 1835–39

5 Right: A positive made from the oldest negative in existence

of light, integral calculus and crystals of borax. His two closest friends at that time were Sir John Herschel and Sir David Brewster, and many well-known scientists such as Roget and Babbage were his acquaintances.

The public announcement of photography

On 7th January, 1839, Daguerre, a French artist, announced his own process for producing photographs, which he called daguerreotypes, but he did not at that time give any details. Talbot, thinking that Daguerre's process must be very similar to his own, on 20th January sent letters to Arago and Biot, two French scientists who had been associated with the announcement, claiming priority for his own discovery. In fact Daguerre's process consisted of exposing to light a film of silver iodide deposited on a silver-plated copper plaque, followed by the treatment with mercury vapour of the faint image produced. The image was then fixed with common salt. It produced positives directly without any intervening negative. Biot replied on 31st January saying that Daguerre had been working on the problem for fourteen years and had produced many processes before the one which he used now. He added, 'The effects it gives excite the admiration of all our artists by their perfection and delicacy'.

Talbot was, no doubt, somewhat annoyed that Daguerre had stolen his thunder and took steps to publicize his own invention. On 25th January, at one of the Friday evening meetings of the Royal Institution, Michael Faraday announced Talbot's new discovery, and examples of photogenic drawings, both negatives and positives, were displayed on the walls of the upper library of the Institution. On 31st January Talbot presented a paper to the Royal Society entitled 'Some Account of the Art of Photogenic Drawing; or the Process by which Nature's Objects May be Made to Delineate Themselves without the Aid of the Artist's Pencil'. The paper gave a brief account of the history of the method and also mentioned some of the uses which Talbot had made of the process. As well as taking views of his house and grounds, he had made shadowgraphs of leaves and flowers (figure 6) and also the first photomicrographs (figure 8) of minute objects enlarged by his solar microscope (figure 7). His exposures for camera views were of the order of one hour. The paper was published in *The Athenaeum* for 9th February.

Sir John Herschel had been unable to attend the Royal Institution and the Royal Society meetings owing to a 'rheumatic affection', but he had heard of Daguerre's and Talbot's announcements. Towards the end of January he carried out some brilliant work of his own. Although no details of either discovery had been published, in a few days he succeeded in producing negative photogenic drawings on paper with silver salts using sodium thiosulphate ('hypo') to remove the unaltered silver salts and so fix the negatives. Herschel had discovered in 1819 that hypo had the property of dissolving silver salts and the fact was included in the standard textbook of the day, W. T. Brande's *Manual of Chemistry*. Talbot used Brande (the book is

6 Right: A 'shadowgraph' made by placing a leaf frond on sensitized paper and exposing to light, 1836

7 Far right: Talbot's solar microscope

8 Right: Photomicrograph of insect wings made with the solar microscope c.1841

referred to in his notebooks) but had not tried hypo for fixing, possibly because it was not readily available. Herschel invited Talbot to his house at Slough to show him this new process of fixing or 'washing out' as Herschel called it. The visit is recorded in Herschel's notebook as follows: 'Friday, Feb. 1. Mr Talbot came to Slough . . . Explained to him all my processes—He also showed me his specimens of results but did not explain his process of what he called 'fixing'—By way of trial of the power of my process of 'washing out' he gave me one of his unfixed specimens. In two minutes I brought him half of it washed the other not and on exposing both the washed half unchanged and the other speedily obliterated and at length grew quite dark'. The two halves are still preserved in the notebook.

During February 1839 Talbot tried several fixatives including hypo but from this date until about April 1842 he continued to fix his photographs with common salt, potassium bromide, or potassium iodide, salts which left silver halides on the paper. The reason why Talbot was unimpressed by the use of hypo may be contained in an entry in Herschel's notebook dated 25th March, 1839: 'Talbot's sensitive paper will not fix well. It is too full of silver for the Hyposuli. soda [hypo]. But the H.S.i. Ammonia will fix it—as the H.S.Ammonia and silver is excessively soluble which that of Soda and silver is not.'

Herschel's photogenic drawings from this period are even more faded than Talbot's, apparently because he did not realise the necessity of washing away all traces of hypo from them after fixing.

At the beginning of May Herschel went to Paris. Any illusions Talbot may have had that his photogenic drawings were a match for the daguerreotype must have been shattered by the letter Herschel wrote to him on 9th May 1839. Herschel described the daguerreotypes he had seen as follows:

> 'It is hardly saying too much to call them miraculous. Certainly they surpass anything I could have conceived as within the bounds of reasonable expectation. The most elaborate engraving falls far short of the richness and delicacy of execution every gradation of light and shade is given with a softness and fidelity which sets all painting at an immeasurable distance. His times also are very short—In a bright day 3m. suffices. . . . In short if you have a few days at your disposition I cannot commend you better than to come and see. Excuse this ebullition. P.S. The pictures are on very thin sheets of plated copper, neither expensive nor very cumbersome.'

Daguerre announced full details of the daguerreotype process in August 1839 and it was demonstrated in London the following month.

Herschel continued his photographic experiments and wrote to Talbot on 10th September 1839: 'I have not tried Daguerre's process—but I yesterday succeeded in producing a photograph on glass [figure 9] having much the character of his results. . . . The process is delicate and very liable to accident in the manipulation, as it consists of depositing on the glass a perfectly uniform film of muriate of silver (silver chloride).' The use of glass as a support instead of paper was to prove very important later when glass negatives eventually superseded the daguerreotype. However, because of difficulties of manipulation both Talbot and Herschel discarded glass as a support until much later.

Talbot worked hard at his photographic experiments during 1839 and 1840. Before March 1839 he had used common salt and silver nitrate to prepare his photogenic drawing paper, but in that month he found that potassium bromide used instead of common salt gave a more sensitive paper. He also turned to larger cameras (figure 10). They remained simply boxes fitted with lenses, but he added a hole above and to one side of the lens, which could be used for focusing the image on the paper or to observe how the paper was darkening. During the exposure the hole was closed, usually with a cork, although according to his notebook it had occurred to him to use

9 Positive from the glass negative made by Herschel on 9th September, 1839

10 Three of Talbot's cameras, 1840–42

red glass for this purpose. Although Talbot made no fundamental advance in the method of taking photographs between March 1839 and September 1840, at least he improved his technique and the appearance of his results. 'I am very much obliged indeed by your very very beautiful photographs', Herschel wrote on 19th June 1840: 'It is quite delightful to see the art grow under your hands in this way. Had you suddenly a twelvemonth ago been shown them how you would have jumped and clapped hands (i.e. if you ever do such a thing).'

Before leaving the subject of photogenic drawings we must mention that the Rev J. B. Reade had also independently discovered the method of producing fixed photographs by a process similar to Talbot's. The date of his discovery is uncertain as he took no steps to publicize it and appears to have had no idea of its potential. No example of his photographs exists. Nevertheless he did have an important influence on the course of photography, as we shall see shortly. The photogenic drawing process was not much used by anyone other than Talbot, Herschel and Reade. Even Sir David Brewster, who received many specimens from Talbot, never tried it out. The process was certainly not a match for the daguerreotype and would have rapidly become forgotten if it were not for the events described in the next section.

2 The discovery of the calotype process

The method of making photogenic drawings, as described in the last section, consisted of exposing sensitized paper in a camera until an image appeared and then fixing the image. No one at that time knew that a much briefer exposure produced what we now call the latent image, which could be developed with a suitable reducing agent. Talbot discovered the latent image and the possibility of developing it in September 1840. The reducing agent he used was gallic acid.

Talbot first used gallic acid in April 1839. The notebook entry reads 'Dilute gallic acid, and dilute nit. silver mixed turn dark in daylight (I believe Mr Reade discovered this)'. The Rev J. B. Reade had in fact discovered the accelerating action which gallic acid had on the speed of photogenic drawing paper and Talbot had heard of this discovery through their mutual acquaintance, Andrew Ross, an optician who was supplying both of them with lenses.

Talbot's experience with gallic acid in April 1839 was not encouraging; but in September 1840 he returned to it to make 'an exciting liquid' by mixing it with a solution of silver nitrate and acetic acid. This liquid was then used to sensitize photogenic drawing paper. On 23rd September 1840 he records what to-day would be described as a great 'break-through'—the development of an image on a paper apparently blank when taken from the camera. The notebook entry reads: 'The same exciting liquid was diluted with an equal bulk of water, and some very remarkable effects were obtained. Half a minute suffices for the Camera. The paper when removed is often perfectly blank but when kept in the dark the picture begins to appear spontaneously, and keeps improving for several minutes, after which it should be washed and fixed with iod. pot. . . . The same exciting liquid restores or revives old pictures on W. paper which have worn out, or become too faint to give any more copies. Altho' they are apparently reduced to the state of yellow iodide [of] silver of uniform tint, yet there is really a difference and a kind of latent picture which may be thus brought out.'

This discovery reduced the exposure time needed from about one hour to one of a few minutes. The new process, which Talbot called the Calotype process (from the Greek καλός beautiful) but which his friends soon called the Talbotype process, consisted of sensitizing a sheet of paper with gallo-nitrate of silver (gallic acid, silver nitrate and acetic acid). The paper was then exposed to light in a camera for one to three minutes and the apparently blank sheet was developed by a further treatment with gallo-nitrate of silver. The resulting negative was then treated with

11 The coachman at Lacock Abbey, 1840. One of the earliest of Talbot's calotypes to include a human figure. The exposure was three minutes

12 Calotype by Talbot taken with an exposure of one minute. The photograph is dated at the lower right

fixative. Positives were still obtained from the negatives by the photogenic drawing process. In appearance the new calotypes were identical with photogenic drawings, but the reduced exposure required widened the scope of photography on paper and paper portraits were possible for the first time (figure 11).

Surprisingly Herschel was not informed of the discovery until five months later. 'I really cannot express the surprise and delight.' Herschel wrote on 16th March 1841, 'with which I read your Circular received this morning giving an account of the Kalotype* (to which I doubt not all the rest of the world will assign the name of Talbotype). I always felt sure you would perfect your processes till they equalled, or surpassed Daguerre's, but this is really magical. Surely you deal with the naughty one'.

*Elsewhere the word was always spelt 'Calotype'

In March 1841, Talbot sent a calotype (figure 12), taken with an exposure of exactly one minute, to Biot, who showed it at a meeting of the Académie des Sciences. Biot remarked that the ivy appeared too dark because of the insensitivity of the salts to green rays and commented that the grain of the paper presented insuperable difficulties to the production of the sharp details which could be obtained on Daguerre's polished silver plates. Talbot had not sent any details of his method, for Biot, in answer to Talbot's letter, wrote: 'I believe I have discovered from your drawing an indication which may well disclose some pecularity of your method of operation. It is the date 23rd February 1841 which I find written unreversed*. As it is without doubt not marked out on the ivy in reversed letters this suggests that, like your first process, there are two operations, one after the other, the one giving the drawing with white instead of black and the second which restores them to their true place by transfer.'

*A direct positive would have given a mirror image reversed with regard to left and right

Daguerre had received a considerable sum from the French Government for the invention of the daguerreotype. He had also patented the invention in England and was receiving further sums of money from this source. Talbot suspected that he would get nothing from the British Government for his invention and, in February 1841, applied for a patent for the calotype process in England. He wrote to Brewster and Herschel to get their opinions. Brewster replied that he was glad Talbot had taken out a patent but an extension of it to Scotland would be unprofitable.

Herschel, we might think, would be against the patent. About a year earlier he had written to Talbot, 'I protest against people taking out patents for any little step they may make in the application of this or any other art'. But there is no disapproval in his reply to Talbot. 'You are quite right in patenting the calotype', he wrote, 'With the liberal interpretation you propose in exercising the patent right no one can complain. And I must say I have never heard of a more promising subject for a lucrative patent of which I heartily give you joy.' Doubtless Talbot had assured Herschel that he would not interfere with the use of the calotype in the sciences and possibly he added that he would be lenient in the enforcement of the patent with regard to amateurs. Unfortunately once Talbot had sold one licence to practise the calotype he was obliged to prosecute anyone else who used the process without a licence, in order to protect the licensee. In fact there was to be considerable complaint in later years.

Talbot, the photographer

Photography was Talbot's recreation as well as his work. Some of the most pleasing of the calotypes acquired by the Science Museum from Lacock Abbey are portraits of Talbot's family, friends and servants taken for his private enjoyment (e.g. figures 13–20). They may be considered as photographs from the first of the many Victorian family albums. Although the exposures must have been of the order of minutes and may well have been an ordeal for the subject, they have an informality lacking in the studio portraits of the day. Talbot seemed keen to take 'simulated instantaneous' photographs of people doing everyday things, like sawing wood or playing the harp, and succeeded so well that one tends to forget the length of exposure. Even at this early stage we see the desire to capture an instant of time on paper. This interest later resulted in the foundation of high-speed photography, when in 1851 Talbot used an electric flash lasting for about 1/100,000th of a second to record the clear image of a page of a rapidly revolving newspaper.

Talbot lived at Lacock Abbey during the 1840s with his wife Constance, his four children, Ela, Rosamund, Matilda and Charles, his mother, Lady Fielding, and his half sister, Miss Horatia Fielding. His stepfather, Admiral Fielding, had died in 1837. There was also a French governess for the children and numerous servants (even the governess had her own maid). Among the frequent visitors was the celebrated Irish poet, Thomas Moore, and his wife Bessy who lived at Sloperton within walking distance of the Abbey.

Moore has written in his *Memoirs:* 'August 18, 1841 . . . started for Lacock Abbey this morning on my way to town. The day beautiful, and I found grouped upon the grass before the house Kit Talbot*, Lady E. Fielding, Lady Charlotte and Mrs Talbot for the purpose of being photogenized by Henry Talbot who was busily preparing his apparatus. Walked alone for a while, about the gardens, and then rejoined the party to see the result of the operation. But the portraits had not turned out satisfactorily, nor (oddly enough) were they at all like, whereas a dead likeness is, in general, the sure, though frightful result of Daguerre's process.' During Moore's walk it is unlikely that anything more than a negative could have been produced which may explain why Moore (who was acquainted with daguerreotypes but may have been seeing the calotype process for the first time) was surprised that a likeness was not obtained.

*Talbot's cousin Christopher

The collection acquired from Lacock Abbey also contains hundreds of calotypes of marine subjects, buildings and landscapes, possibly the work of several photographers. Reproductions of some of these have been included (figures 21–25). Two of the most interesting are of Nelson's Column in the course of construction (figure 24) and Hungerford Suspension Bridge (figure 25). This bridge was designed by Brunel and was opened as a toll bridge in 1845. It was taken down in 1862 to make way for the present Charing Cross Railway Bridge.

13 The Rev C. R. Jones with two others

14 A Calotype portrait being taken at Lacock Abbey c.1842

15 The Keeper at Lacock Abbey

16 A group at Lacock Abbey

18 Talbot's three daughters, Ela (b.1835), Rosamund (b.1837) and Matilda (b.1841)

17 Unknown lady

19 Miss Horatia Fielding, Talbot's half-sister

20 Calotype portrait

21 Sailing ship being careened

22 Left: Salt beds, possibly at Malta

23 Opposite: Sailing ships at Swansea

24 Above: Trafalgar Square, c.1844

25 Hungerford Suspension Bridge, c.1845

3 The exploitation of the calotype process

Henry Collen, the first licensee

1841 was the year in which photography was introduced to towns in Britain in the form of daguerreotype portrait studios. By the end of the year both Beard and Claudet had established studios in London, there was a studio on the Marine Parade, Brighton, and several other studios were operating in both the north and south of England. The early daguerreotypists were making small fortunes, though the main attraction of the early portraits was their novelty rather than their beauty, and they received several unfavourable notices in the press. One such notice appeared in *Blackwood's Magazine* in 1842: 'The likenesses produced, by the (daguerreotype) system, are so absolutely fearful, that we have but little hope of ever seeing anything tolerable from any machine. It must want colour, it must want living expression, it must want the play of features, which the pencil has the singular power of seizing and fixing; and its best likeness can be only that of a rigid bust, or a corpse.'*

**Blackwoods Magazine* 517 (1842)

The most displeasing feature of the early daguerreotype portraits was their clinical, cold, metallic silver surface, later alleviated by toning and colouring. Calotype portraits, made possible by the discovery of the development process, were in warm tones, and had the additional advantage that, in contrast to the daguerreotype, their surfaces were not fragile and so could be retouched. The first press notices of calotype portraits appeared in the spring of 1842 and were favourable. The *Morning Post* said, 'The portraits, those at least we have seen, are very satisfactory. There is a rough air of truth about them, which reminds one of the first, and sometimes the best, sketches of the artist—a sort of free sepia, or, rather, lithotint drawing, full of broad effects and vigour.'

As calotype portraits could be retouched it is not surprising that the first professional calotypist was a painter of miniatures, Henry Collen. Calotype portraiture was barely possible with the conventional lenses of maximum aperture f15, the exposure required at this aperture being of the order of several minutes. To reduce the exposure time Collen approached the leading optical lens maker in London, Andrew Ross, to obtain a lens of larger aperture. The lens Ross produced is now in the Royal Photographic Society's collection and is exhibited at the Science Museum. To all outward appearance it is a symmetrical doublet, but if we look at the internal construction (figure 26) we see that the two compound lenses which make up the doublet differ internally. The lens is of eleven inches focal length and of maximum aperture about f4. At large apertures, because of the large field curvature of

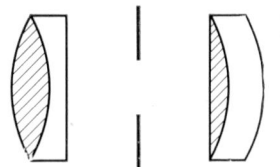

26 Diagram of the lens made by Andrew Ross for Henry Collen

27 Dr Boyd, a specialist in mental disorders. A retouched portrait by Henry Collen

the lens, Collen had to resort to the use of curved pieces of ground-glass 8 inches in diameter for focussing and to similar pieces of plain glass to hold the sensitive paper in position during the exposure. At smaller apertures flat glass was used.

The earliest of Collen's portraits were produced in the summer of 1841 and were processed by Talbot, who appears to have been very closely connected with the venture. Collen retouched so extensively (figure 27) that he may have considered the photographic portrait only a guide for his pen. He sought to combine the charm of an artist's sketch with the accuracy of a photograph. To-day the photographs are faded, leaving the lines of the pen more prominent, and we need to use a little imagination to judge his degree of success. Sir David Brewster thought highly of them. On 22nd March 1842, he wrote to Talbot, 'Mr Collen has been so kind as to send me one of his calotypes which has astonished me and all who have seen it. Dr Adamson to whom I have just shown it despairs of ever coming near it.'

Collen produced about one thousand of these photographic sketches but only worked with the process for about a year. We do not know when Collen's studio closed but we do know that by the autumn of 1842 Talbot was seeking another partner to take up the calotype professionally in London and got in touch with Antoine Claudet.

Antoine Claudet

Claudet had opened the second daguerreotype portrait studio in London—at the Adelaide Gallery in the Strand (the first was opened by Richard Beard). Claudet's early daguerreotype portraits were no better than those of the other operators and he later admitted that he was ashamed of them. But he was a man of learning and a scientist and he strove to improve the daguerreotype. By 1844 he had done much to rescue its reputation from the depths to which it had sunk in 1842. He made several contributions to photographic science and became, in 1853, a Fellow of the Royal Society. It is little wonder that Talbot had more affinity with him than with any other of the early daguerreotypists.

In October 1842, Talbot inquired of Claudet whether he could be persuaded to practise the calotype at the Adelaide Gallery. Claudet was quite prepared to collaborate on the project but negotiations dragged on for eighteen months. The main bone of contention was a clause which Talbot wanted to insert into the agreement to the effect that if Talbot, who was to receive 25% of the takings of the calotype operations, received less than £400 in one year he was to be free to grant other licences in London. Claudet maintained that this should not apply during the first two years of the agreement when he would be perfecting the technique of using the new process and would be involved in the expense of bringing it before the public. The price of portraits was mutually agreed at one guinea for a single print reducing to five shillings for copies. (This was at a time when labourer's wages might be twelve shillings a week and a foreman's wage perhaps double, so portraits were only for the more wealthy). Claudet thought his expenses for running a calotype studio might be £30 a week, including £8 to £10 a week for advertising, and so he said he would need to take five portraits a day to cover his expenses. However, by the summer of 1844 an agreement had been signed and Claudet began to receive instruction in the calotype process from Talbot's assistant Nicholaas Henneman (figure 28).

28 Nicholaas Henneman holding a copy of *Pencil of Nature*. Calotype, 1844

At the end of August Claudet was able to send some examples of his calotype work for Talbot's inspection and included a self-portrait (figure 29). Another of the calotypes which Talbot received from Claudet was of the British consul George Pritchard (figure 30). Pritchard had been the consul on Tahiti but was expelled as a result of French military action in May 1844. He arrived in London during the summer and his exploits were recorded in the newspapers. 'He came to have made a daguerreotype', wrote Claudet, 'and we took advantage of the opportunity to make a Talbotype. The negative is very beautiful and we have made three beautiful copies. I believe you will be satisfied.'

However, it was not all plain sailing for Claudet. One of the most awkward of his early customers was Stephen Spring-Rice, a Commissioner of Customs and an influential person. 'Mr Stephen Spring-Rice came to have his portrait taken by the Talbotype,' wrote Claudet (23rd August 1844). 'We have attempted it three times. The first time the light was bright and he was unable to put up with the sun. We waited some time for a cloud and then we operated with a pose of twenty-five seconds. As there were spots on the negative I invited Mr Spring-Rice to return. You will see the two negatives from the subsequent operations. They were made when the weather was dull. To-day to our great astonishment I received a letter in which he says, "Mr Spring-Rice, having received a letter from Mr Fox Talbot stating that nothing is easier than to obtain a good portrait by his process in 5 or 6 seconds, Mr Spring-Rice thinks that the most favourable interpretation of the repeated failures at the Adelaide Gallery is that the operator does not understand the process. He will not therefore trouble that person to make another attempt." I leave you to judge our failure.' The exposure of twenty-five seconds mentioned here is one of the shortest calotype exposures recorded. Often the exposures were over a minute.

Claudet was concerned lest Talbot should infer from this incident that he was neglecting the calotype. 'I assure you it is not so', he wrote the following day, 'and that I spare no expense or trouble to make it a success. I would be ashamed if anyone could accuse me of the contrary. . . . Until we have a paper with a surface as uniform and perfect as a silver plate I say that the Daguerreotype gives images more delicate, finer and of greater perfection than the Talbotype. Until we can operate with the Talbotype in several seconds and as rapidly with as the Daguerreotype so that one can get more pleasing poses, then I say that the advantage is on the side of the Daguerreotype. But I also say that the Talbotype has beauty which the other has not, that the impressions are more portable and circulate more easily, that it is possible to send them through the post, stick them in albums, etc, and finally one can obtain an unlimited number of copies.'

Claudet favoured the large portrait of whole-plate size for the calotype, for then the grain of the paper was less noticeable. He persevered with the calotype process but with limited success. By November 1844 he had facilities for processing forty negatives a day 'but it will be a long time before we reach this number'. In fact

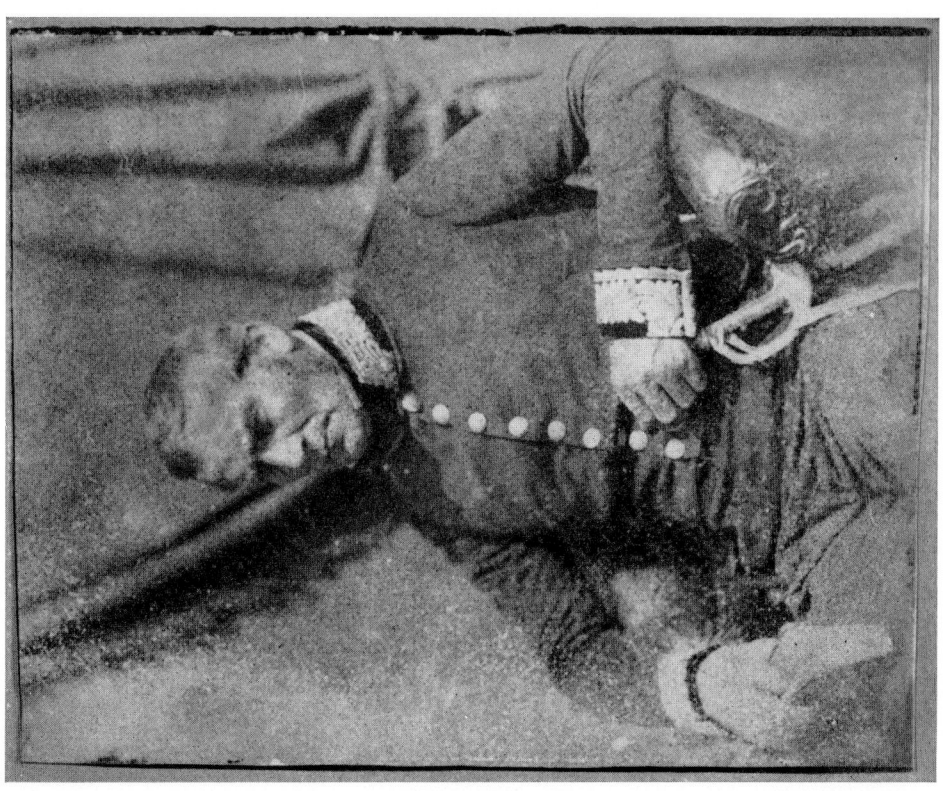

30 George Pritchard in consul's uniform. Calotype by Claudet, August 1844

29 Antoine Claudet. Calotype self-portrait. August 1844

profits from the sale of calotype portraits were disappointing and he returned to the daguerreotype which was more reliable and appears to have had a greater appeal to the public.

However, Claudet continued to hold Talbot in high regard. In 1853, when Claudet was planning the decoration of the reception room of his new photographic establishment in Regent Street, painted portraits of Daguerre and Talbot were together given pride of place, with portraits of other inventors who had contributed to the science of photography grouped on either side.

31 Talbot's printing establishment at Reading

The printing establishment at Reading

Undoubtedly the main advantage of the calotype compared with the daguerreotype was the ease with which a large number of positives could be produced from a single negative. Talbot was very conscious of this and sometime around the beginning of 1844, he set up a printing establishment at Reading, the main purpose of which was to produce a very large number of prints for publication.

The man Talbot put in charge of the photographic side of the Reading establishment was Nicholaas Henneman. Henneman had been Talbot's valet and assistant since the 1820s and had a great respect for him. Talbot trusted him to the extent of allowing him to recruit staff and suggest wages at Reading. But Henneman was not in sole charge. His English spelling and grammar were both poor (he was a Dutchman) and so it is unlikely that Talbot would have been happy to have him writing letters to clients. For this side of the business Talbot appointed B. Cowderoy, a man of whom little is known except that he was well educated and dealt with applications for licences and sets of calotypes.

During the three to four years that the establishment was in operation tens of thousands of calotypes were produced there, many for sale in stationers' shops up and down the country at a price of one to five shillings depending on size. We see some of these calotypes being taken in the well-known photographs of the Reading establishment (figure 31). The subjects of the calotypes produced for sale were mainly buildings, engravings and statuettes.

It was at Reading that the calotypes used in the first book to be illustrated with actual photographs, *Pencil of Nature*, were produced. The photographs were simply gummed into the leaves of the book. *Pencil of Nature* was published in six parts between June 1844 and April 1846. The complete work contained 24 calotypes of approximately whole-plate size. Talbot included a wide variety of calotypes in the book but did not include a portrait. In the text which accompanied the calotypes he gave a brief account of the history of the invention and some idea of its applications. A second publication, *Sun Pictures in Scotland*, containing 23 calotypes was produced in 1845.

Another mammoth effort at publicizing the calotype was carried out in 1846, when each issue of the magazine *Art Union* for June of that year included a whole-plate calotype. The calotypes were made from a wide assortment of negatives. Many thousands were produced for this issue but, as we shall see later, the project was not an unqualified success. The last of the major commissions of the Reading establishment was the production of calotypes for Sir William Stirling's *Annals of the Artists of Spain*. This again involved making over fifteen hundred calotypes. By May 1847, Henneman had completed Sir William Stirling's order and soon afterwards the establishment closed.

Shortly after the Reading establishment closed Talbot set up a calotype studio of his own at 122 Regent Street. In charge of the studio were Nicholaas Henneman and Thomas A. Malone, a man whom Henneman had recruited in the summer of 1844, but who had more knowledge of chemistry than Henneman himself.

At first, trade at the studio was slow. It had been found at Reading that the calotypes produced tended to fade, and the ones produced at London were even more prone to this defect. The fact was becoming generally known. In July 1849, Malone wrote to Talbot, 'Mr A. Taylor of Guy's Hospital has spoken of the fading publicly

*Vernon Heath later became a professional photographer

at the Royal Institution and at Lord Rosse's Soiree, where our pictures were exhibited and much admired. A scientific amateur Mr Vernon Heath . . . lately sold his camera 'in disgust' because he could not fix his sun pictures.* A clergyman taught by us also complains that after all the care he has bestowed upon his pictures they fade. . . . We are constantly and unpleasantly cross-examined on the subject. The Art Union copies containing sun pictures seem to have done harm. The artists who have them are interested in giving every publicity to the fact of their growing faintness.'

Malone was a capable chemist and quite rightly attributed the fading to traces of hypo remaining in the calotypes. To enable him to devote more time to work on a new fixing process Malone appointed a Herr Kahn to help Henneman with the portraiture, 'but his pictures are not preferred to our own' and Herr Kahn left a month later. Malone did produce a new fixing process but Talbot was sceptical of it and said that he would have to wait a year to see how well the new photographs lasted. Once a new and reliable fixative had been found he intended to go ahead and produce further issues of *Pencil of Nature*.

Later apparently business at the Regent Street studio improved, for in 1852 Henneman purchased a property in Kensal Green which was used for printing positives.

The introduction of negative-positive photography to Scotland

As already mentioned one of Talbot's best friends was Sir David Brewster (figure 32) who was, from 1838, Principal of the University of St Andrews.

32 Sir David Brewster

At the beginning of 1839, Talbot sent Brewster examples of photogenic drawings which Brewster showed to his university colleagues. By the following January Brewster was asking for precise details of the method, for 'I believe I possess everything you have written on the subject and yet I feel that I could not produce anything like your present specimens'. Within a few weeks of the discovery of the possibility of development of the latent image in September 1840, Talbot had written to Brewster to tell him of his new process. He did not send any details of the chemicals which brought about the development.

By this time Brewster had acquired a daguerreotype camera and several times wrote to Talbot for details of the calotype process. Talbot was loathe to let the secret out until he had secured it with a patent, and did not send details until May 1841. He also sent some gallic acid which was unobtainable in Edinburgh. Even then Talbot did not send any information on how to make a positive from a negative. 'Of course', Brewster wrote, 'this is done by superposition; but your experience must have taught you the best way of doing it. I cannot understand how anything so sharp as the two Positive Portraits you sent me can be produced by the Solar Rays after they have permeated, and been to a certain degree dispersed by a piece of writing-paper.' Even with the manipulative details Brewster entirely failed to obtain calotypes. He had also given the details to his two friends at St Andrews, Major Playfair and Dr John Adamson. 'Dr Adamson, who is a good Chemist and successful with the Daguerreotype has also failed, and says that the paper when ready for the camera became black in the dark' (July 1841). The first partial success came in November 1841, when Brewster was able to send Talbot three of Dr Adamson's calotypes fixed with common salt or potassium bromide. Major Playfair, however, was still unsuccessful, only getting faint images after exposure of four minutes. By March 1842, Talbot was very pleased with Dr Adamson's results.

During the first half of 1842, Talbot wrote to Brewster asking him to see if he could induce anyone to practise calotype photography as a profession in Scotland. In August 1842, we first hear of Dr Adamson's brother Robert. 'A brother of Dr Adamson . . . is willing to practise the calotype in Edinburgh as a Profession' wrote Brewster. 'Mr Adamson . . . has been well drilled in the new art by his brother'. By the middle of 1843, Robert Adamson was established in Edinburgh 'with crowds every day at his studio'.

In May 1843, there took place in Edinburgh the First General Assembly of the Free Protesting Church of Scotland. An artist had decided to paint a picture representing the whole assembly of about 500 ministers. Brewster wrote (3rd July 1843) 'I got hold of the artist—showed him your calotype and the eminent advantage he might desire from it in getting likenesses of all the principal characters before they dispersed to their respective homes. He was at first incredulous, but went to Mr Adamson, and arranged with him the preliminaries for getting all the necessary Portraits. They have succeeded beyond their most sanguine expectations—They

33 The Rev James Fairbairn and Newhaven fishwives. Calotype by Adamson and Hill, autumn 1843

have taken, on a small scale, groups of 25 persons in the same picture all placed in attitudes which the painter desired, and very large pictures besides have been taken of each individual to assist the Painter in the completion of his picture. Mr D. O. Hill the painter is in the act of entering into partnership with Mr Adamson and proposes to apply the Calotype to many other general purposes of a very popular kind. . . . I think you will find we have, in Scotland, found out the value of your invention not before yourself, but before those on whom you have given the privilege of using it . . . The Daguerreotype is considered infinitely inferior, for all practical purposes, notwithstanding its beauty and sharpness.'

The collaboration of Robert Adamson and D. O. Hill lasted for about five years during which time they produced about fifteen hundred pictures. It seems a reasonable assumption that Hill never took the trouble to master the manipulative technique of calotyping, for when Adamson died in 1848 he abandoned photography and went back to painting.

Hill had access to the highest social circles in Edinburgh and the Adamson and Hill calotypes mainly consist of portraits of celebrities. But they also visited fishing villages on the East coast of Scotland and took calotypes of fishermen and fishwives (figure 33), which Brewster particularly liked.

The calotypes of Adamson and Hill are now considered by many to be among the masterpieces of photography and they are given the credit of being the first photographers to indicate the artistic potential of the medium.

The calotype in America

The daguerreotype was practised to a wider extent in America than in any other country. By 1849 there were more than 60 daguerreotype studios in New York, 30 in Philadelphia and similar numbers in all the other big cities. Yet the calotype process was unknown. Talbot had patented the calotype process in America in 1847. The first applicant for the patent rights was Edward Anthony, but two months later he wrote to say that he was no longer interested.

In February 1849, William and Frederick Langenheim, two brothers who had been operating a daguerreotype studio in Philadelphia since June 1841, wrote to Talbot to see if they could persuade him to allow them to be his agents in America. Voigtländer of Vienna was a brother-in-law of the Langenheims and they already had the American agency for his daguerreotype apparatus, of which they had sold the impressive number of 1,500. The terms they offered Talbot seemed at first sight generous. They would receive 25% of the amount of the sales of patent rights, etc and return 75% to Talbot. Talbot, possibly bearing in mind the difficulties of verifying quarterly accounts of agents 3,000 miles away, refused this.

In May 1849, William Langenheim visited Talbot at Lacock Abbey and together they agreed on an outright sum of £1200, to be paid in instalments. Talbot was pleased with the deal. The Langenheim brothers immediately set about publicizing the calotype in the United States. Unfortunately, events were not in their favour. Cholera, blamed on the influx into Canada of destitute Irish escaping from the potato famine in their homeland, was causing panic along the East coast of America. All people of wealth had moved out of the cities into the countryside. Also it was the time of the Californian Gold Rush, and the more enterprising of the young men were going West to seek their fortunes.

By September 1849, it was clear to the Langenheim brothers that they were going to have great difficulty in paying the two remaining £300 instalments of the purchase price of the patent rights. To convince Talbot that they had not been wasting their time they sent him a brochure describing the calotype process (they had distributed a thousand of these brochures) and some of their best calotypes (figure 34). With

34 A lighthouse under construction. Calotype by W. and F. Langenheim, 1849

regard to sharpness of detail these calotypes were superior to Talbot's and rivalled daguerreotypes. The Langenheims had developed a method of making negatives on glass by coating the glass with silver salts in the white of egg (albumen).

By November, notwithstanding strenuous efforts, they had only sold patent rights in five unimportant states (Georgia, Florida, Alabama, Louisiana and Texas). The daguerreotype process was so firmly established along the wealthy East coast that they found it impossible to break into the market. The amount their Philadelphia portrait studio brought in was insufficient to meet their debt. Instead of paying it, the brothers suggested that Talbot should take the patent rights for their albumen-on-glass process, but Talbot refused.

This unsuccessful speculation on the calotype process contributed to the failure of the firm of W. and F. Langenheim in 1851. After the failure Frederick Langenheim spent three years in South America. When he returned to Philadelphia in 1854, he found that portrait studios were turning to the wet collodion process which Talbot claimed was still covered by his patent. Again he approached Talbot to get the patent rights, but at this time Talbot was engaged in the law suit concerning his patent, which will be mentioned later, and the application came to nothing.

The calotype in France

Talbot patented the calotype process in France in the early 1840s.

In 1842 he bought some photographic equipment from Charles Chevalier, a Paris optician, and sent him some calotypes. Chevalier thought the photographs were remarkable. He replied 'your process will without doubt be the most used by artists and travellers. . . . Many people would use photography on paper if they had the necessary instructions but the information in reports that you give is not sufficiently detailed.'

Possibly Talbot believed that the materials were so cheap and the process so easy that once the full details were published anyone could use the process and would not take the trouble to apply for a licence. No full details were published and the only calotype licence sold in France that we know of was to a Maquis de Bassano in 1844.

In 1846 Chevalier told Talbot that the calotype was not much used in Paris as details were still not easily available.

Towards the end of the 1840s, Blanquart-Evrard of Lille started selling calotypes on a large scale in France. His process was a modification of Talbot's with several improvements which made the results less liable to fade and of a more pleasing appearance. Talbot received several letters telling him of the fine calotypes on sale in Paris but took no action against the infringer of his patent. He seemed most annoyed at the fact that the French called them daguerreotypes on paper.

Blanquart-Evrard, like Talbot, believed that the greatest advantage of the calotype process and the greatest potential of photography lay in book illustration. In 1851 he set up a printing establishment at Lille similar to Talbot's Reading establishment, but on an even larger scale. The best known publication of the Lille establishment is *Egypte, Nubie, Palestine et Syrie*, a book containing no less than one hundred and twenty-five calotypes from negatives taken by Maxime du Camp in the Middle East (1849–51).

The end of the calotype process

During the whole of the 1840s the calotype and the daguerreotype processes competed. In 1851 Frederick Scott Archer published a new process, the 'wet collodion' process, which was better than either of them. It was faster and could give as great detail as the daguerreotype and had nearly all the advantages of the calotype. Negatives were taken on glass coated with collodion containing silver iodide and were developed with pyrogallic acid. The glass negatives were then used to make paper positives or, for portraits, to make glass collodion positives. The most serious disadvantage of the new process was that the glass negatives needed to be sensitized immediately before use. Nevertheless it was sufficiently superior to both the calotype and daguerreotype to sound the death-knell of both processes.

Talbot considered that the collodion process was still covered by his patent. The process was a negative-positive one using light-sensitive silver salts to produce a latent image which was then developed in a manner similar to his own calotype process. When used to prepare positives on paper the results were similar. But at that time there was substantial opposition to Talbot's exercise of the patent rights. It was considered that he was impeding the advance of negative-positive photographic processes by the way he interpreted the patent.

In view of this opposition Talbot in 1852 relinquished his patent rights except with regard to professional portraiture. He still regarded the collodion process as covered by his patent and took steps to prohibit its use in professional portrait studios by operators who had not purchased a licence from him. Some of the operators contested his right to exercise the calotype patent over the wet collodion process. A test-case came to the law courts towards the end of 1854 and the trial lasted for two and a half days. The verdict was that Talbot's patent rights did not cover the wet collodion process.

By the middle of the 1850s, both the calotype and the daguerreotype processes were virtually dead. The calotype process was never as extensively practised as the daguerreotype. Yet its rich tones and texture make it the more beautiful process. Because it was the first negative-positive process it is historically more important than the daguerreotype. Talbot had the satisfaction of knowing that by inventing such a process he would always hold first place among photographic scientists.

Select Bibliography

A list of the more important publications of W. H. Fox Talbot relating to his photographic work:

Books

Pencil of Nature (1844) *Sun Pictures in Scotland* (1845).

British Patents

No 8842 (February 8th 1841), No 9753 (June 1st 1843),

No 12,906 (December 19th 1849), No 13,664 (June 12th 1851).

Articles and Papers

Some account of the art of photogenic drawing, or the process by which natural objects may be made to delineate themselves without the aid of the artist's pencil. *Phil. Mag.* **14**, 196–211 (1839).

Remarks on M. Daguerre's photogenic process. *Brit. Assoc. Rep.* 3–5 (1839, part 2).

An account of the processes employed in photogenic drawing. *Roy. Soc. Proc.* **4**, 124–126 (1839).

Note respecting a new kind of sensitive paper, *Roy. Soc. Proc.* **4**, 134 (1839).

On calotype photogenic drawing. *Phil. Mag.* **19**, 88–92 (1841).

An account of some recent improvements in photography. *Roy. Soc. Proc.* **4**, 312–316 (1841).

On the coloured rings produced by iodine on silver, with remarks on the history of photography. *Phil. Mag.* **22**, 94–97 (1843).

On the production of instantaneous photographs. *Phil. Mag.* **3**, 73–77 (1852).

Appendix A of *A History and Handbook of Photography* by G. Tissandier, 345–366 (published posthumously, 1878).

Letters

Literary Gazette, February 27th 1841. *The Times*, August 13th 1852.

Publications on the calotype process by other authors:

Manipulative details for the calotype process were published in many of the early photographic handbooks e.g. Hunt, R. *A Manual of Photography* 4th ed. (1854).

Accounts of the history of the calotype process will be found in all the histories of photography (e.g. Eder, Gernsheim, Newhall, Werge).

The following books and articles deserve note:

Schwarz, H. *David Octavius Hill* (1932).

Johnston, J. Dudley, 'William Henry Fox Talbot' Part 1 (1800–1833) *Phot. J.* **87**, 3–13 (1947)

White, H. *Phot. J.* **89** (Section A), 247–251 (1949).

Newhall, B. 'William Henry Fox Talbot'. *Image* 8, 60–75 (1959).

Gernsheim, H. 'Talbot's and Herschel's Photographic Experiments in 1839'. *Image* **8**, 132–137 (1959).

Dunbar, A. 'The Work of David Octavious Hill, RSA' *Phot. J.* **104**, 53–65 (1964).

Newhall, B. *Latent Image* (1967).

Smith, R. C. 'William Henry Fox Talbot, FRS' *Phot. J.* **108**, 361–371 (1968).

Snow, V. F. and Thomas, D. B. 'The Talbotype Establishment at Reading – 1844 to 1847' *Phot. J.* **106**, 56–67 (1966).

Acknowledgements

The author thanks the Royal Photographic Society for lending the Science Museum the solar microscope shown in figure 8 and some of the cameras shown in figures 5 and 10. Thanks are also due to the National Maritime Museum for permission to publish figure 21.